Lecture Notes in Mathematics

An informal series of special lectures, seminars and reports on mathematical topics

Edited by A. Dold, Heidelberg and B. Eckmann, Zürich

17

Claus Müller

Institut für Reine und Angewandte Mathematik
Technische Hochschule Aachen

Spherical Harmonics

1966

Springer-Verlag · Berlin · Heidelberg · New York

Library of Congress Catalog Card Number 66-22467. Printed in Germany. Title No. 7337.

PREFACE

The subject of these lecture notes is the theory of regular spherical harmonics in any number of dimensions. The approach is such that the two- or three-dimensional problems do not stand out separately. They are on the contrary regarded as special cases of a more general structure. It seems that in this way it is possible to get a better understanding of the basic properties of these functions, which thus appear as extensions of well-known properties of elementary functions. One outstanding result is a proof of the addition theorem of spherical harmonics, which goes back to G. Herglotz. This proof of a fundamental property of the spherical harmonics does not require the use of a special system of coordinates and thus avoids the difficulties of representation, which arise from the singularities of the coordinate system.

The intent of these lectures is to derive as many results as possible solely from the symmetry of the sphere, and to prove the basic properties which are, besides the addition theorem, the representation by a generating function, and the completeness of the entire system.

The representation is self-contained.

This approach to the theory of spherical harmonics was first presented in a series of lectures at the Boeing Scientific Research Laboratories. It has since been slightly modified.

I am grateful to Dr. Theodore Higgins for his assistance in preparing these lecture notes and I should like to thank Dr. Ernest Roetman for a number of suggestions to improve the manuscript.

February 1966 Claus Müller

CONTENTS

General Background and Notation 1

Orthogonal Transformations 5

Addition Theorem 9

Representation Theorem 11

Applications of the Addition Theorem 14

Rodrigues Formula 16

Funk - Hecke Formula 18

Integral Representations of Spherical Harmonic 21

Associated Legendre Functions 22

Properties of the Legendre Functions 29

Differential Equations 37

Expansions in Spherical Harmonics 40

Bibliography 45

GENERAL BACKGROUND AND NOTATION

Let $(x_1,\dots,x_q)$ be Cartesian coordinates of a Euclidean space of q dimensions. Then we have with

$$|x|^2 = r^2 = (x_1)^2 + \dots + (x_q)^2$$

the representation

$$x = r\xi$$

where

$$\xi = (\xi_1,\dots,\xi_q) \text{ and } |\xi| = 1 \qquad 1)$$

represents the system of coordinates of the points on the unit sphere in q dimensions. It will be called Ω_q, its surface element $d\omega_q$ and the total surface ω_q, where this surface is given by

$$\omega_q = \int_{\Omega_q} d\omega_q$$

By definition we set $\omega_1 = 2$. Then we have

$$\omega_2 = 2\pi \; ; \; \omega_3 = 4\pi$$

If the vectors $\varepsilon_1,\dots,\varepsilon_q$ are an orthonormal system, we may represent the points on Ω_q by

$$(1) \qquad \xi_q = t\cdot\varepsilon_q + \sqrt{1-t^2}\,\xi_{q-1} \; ; \; -1 \le t \le 1 \; ; \; t = \varepsilon_q \xi_q$$

where ξ_{q-1} is a unit vector in the space spanned by $\varepsilon_1,\dots,\varepsilon_{q-1}$
The surface element of the unit sphere then can be written as

$$d\omega_q = (1-t^2)^{\frac{q-3}{2}}\, dt\, d\omega_{q-1}$$

and we have from above

$$\omega_q = \int_{\Omega_{q-1}} \int_{-1}^{+1} (1-t^2)^{\frac{q-3}{2}}\, dt\, d\omega_{q-1} \,.$$

The integral on the right hand side may be transformed to

$$\int_0^1 (1-u)^{\frac{q-3}{2}}\, u^{-1/2}\, du = \frac{\Gamma(\frac{1}{2})\,\Gamma(\frac{q-1}{2})}{\Gamma(\frac{q}{2})}$$

---

1) Here and in the following points of the unit sphere are denoted by greek letters.

Which gives us for $q = 2,3,\dots$

(2) $$\omega_q = \frac{\sqrt{\pi}\,\Gamma(\frac{q-1}{2})}{\Gamma(\frac{q}{2})}\,\omega_{q-1} = \frac{\sqrt{\pi}^{\,q-1}\,\Gamma(\frac{1}{2})}{\Gamma(\frac{q}{2})}\,\omega_1 = \frac{2(\pi)^{\frac{q}{2}}}{\Gamma(\frac{q}{2})}$$

Denote by

(3) $$\Delta_q = \left(\frac{\partial}{\partial x_1}\right)^2 + \left(\frac{\partial}{\partial x_2}\right)^2 + \dots\dots + \left(\frac{\partial}{\partial x_q}\right)^2$$

the Laplace operator. We then introduce the

Definition 1 : Let $H_n(x)$ be a homogeneous polynomial of degree n in q dimensions, which satisfies

$$\Delta_q\, H_n(x) = 0$$

Then

$$S_n(\xi) = \frac{1}{r^n}\, H_n(r\xi) = H_n(\xi)$$

is called a (regular) spherical harmonic of order n in q dimensions.

From this we get immediately

Lemma 1 : $$S_n(-\xi) = (-1)^n\, S_n(\xi)$$

Let $H_n(x)$ and $H_m(x)$ be two homogeneous harmonic polynomials of degree n and m. Then by Green's theorem we have

$$0 = \int\limits_{|x|\leq 1} (H_n\,\Delta_q H_m - H_m\,\Delta_q H_n)\, dV = \int\limits_{\Omega_q} H_n(\xi)\, H_m(\xi)\,(m-n)\, d\omega_q\,,$$

as the normal derivatives of H_m and H_n on Ω_q are

$$\left\{\frac{\partial}{\partial r} H_m(r\xi)\right\}_{r=1} = m\, H_m(\xi) \quad \text{and} \quad \left\{\frac{\partial}{\partial r} H_n(r\xi)\right\}_{r=1} = n\, H_n(\xi) \quad \text{respectively.}$$

From Definition (1) we have therefore

Lemma 2 : $$\int\limits_{\Omega_q} S_n(\xi)\, S_m(\xi)\, d\omega_q = 0 \quad \text{for} \quad m \neq n$$

Any homogeneous polynomial in q variables can be represented in the form

(4) $$\sum_{j=0}^{n} (x_q)^j A_{n-j}(x_1,\dots,x_{q-1}) = H_n(x)$$

where the $A_{n-j}(x_1,\dots,x_{q-1})$ are homogeneous polynomials of degree $(n-j)$ in $x_1,\dots,x_{q-1}$. Application of the Laplace operator in the form

$$\Delta_q = \left(\frac{\partial}{\partial x_q}\right)^2 + \Delta_{q-1}$$

gives

$$\Delta_q H_n(x) = \sum_{j=2}^{n} j(j-1)(x_q)^{j-2} A_{n-j} + \sum_{j=0}^{n-2} (x_q)^j \Delta_{q-1} A_{n-j} .$$

For a harmonic polynomial this has to vanish identically. By equating coefficients we thus get

(5) $$\Delta_{q-1} A_{n-j} = -(j+2)(j+1) A_{n-j-2}$$

Therefore all the polynomials A_j are determined if we know A_n and A_{n-1}. The number of linearly independent homogeneous and harmonic polynomials is thus equal to the number of coefficients of A_n and A_{n-1}.

Denote by $M(q,n)$ the number of coefficients in a homogeneous polynomial of degree n and q variables. It then follows from (4) that

(6) $$M(q,n) = \begin{cases} \sum_{j=0}^{n} M(q-1,j) & , n \geq 0 \\ 0 & , n < 0 \end{cases}$$

Clearly $M(1,n) = 1$, so that $M(q,n) = \mathcal{O}(n^{q-1})$.

Now the total number of coefficients available in $A_n(x_1,\dots,x_{q-1})$ and $A_{n-1}(x_1,\dots,x_{q-1})$ is

(7) $$N(q,n) = M(q-1,n) + M(q-1,n-1) = \mathcal{O}(n^{q-2}) \quad , (n \geq 1)$$

Then the power series

(8) $$g_q(x) = \sum_{j=0}^{\infty} x^j N(q,j)$$

converges for $|x| \leq 1$. By (6) and (7)

(9) $$N(q,n) = \sum_{j=0}^{n} N(q-1,j), \quad (n \geq 1).$$

Now it follows from (7)

$$N(1,n) = \begin{cases} 1 & \text{for } n = 0,1 \\ 0 & \text{for } n > 1, \end{cases}$$

so that

$$g_1(x) = 1+x .$$

Substituting (9) into (8) and interchanging the order of summation we obtain

$$g_q(x) = \frac{1}{1-x} g_{q-1}(x)$$

and hence

$$g_q(x) = \frac{1+x}{(1-x)^{q-1}} .$$

This gives us

Lemma 3 : The number N(q,n) of linearly independent spherical harmonics of degree n is given by the power series

$$\frac{1+x}{(1-x)^{q-1}} = \sum_{n=0}^{\infty} N(q,n)\, x^n .$$

Specializing to q = 2 and q = 3 we get

(10)
$$\frac{1+x}{1-x} = 1 + \sum_{n=0}^{\infty} 2x^n = \sum_{n=0}^{\infty} N(2,n)\, x^n$$
$$\frac{1+x}{(1-x)^2} = \sum_{n=0}^{\infty} (2n+1)\, x^n = \sum_{n=0}^{\infty} N(3,n)\, x^n .$$

From Lemma 3 we can determine the N(q,n) explicitly. The binomial expansion gives for $|x| < 1$

$$\frac{1+x}{(1-x)^{q-1}} = (1+x) \sum_{n=0}^{\infty} \frac{\Gamma(n+q-1)}{\Gamma(n+1)\cdot\Gamma(q-1)}\, x^n$$

$$= 1 + \sum_{n=1}^{\infty} \frac{(2n+q-2)\,\Gamma(n+q-2)}{\Gamma(n+1)\cdot\Gamma(q-1)}\, x^n$$

so that

(11) $$N(q,n) = \begin{cases} \dfrac{(2n+q-2)\ \Gamma(n+q-2)}{\Gamma(n+1)\ \Gamma(q-1)}, & n \geq 1 \\ 1, & n = 0 \end{cases}$$

If we set

(12) $$S_n(\xi) = \sum_{j=1}^{N(q,n)} c_j^n \, S_{n,j}(\xi)$$

we have

<u>Lemma 4</u>: There exist N(q,n) linearly independent spherical harmonics $S_{n,j}(\xi)$ of degree n in q dimensions and every spherical harmonic of degree n can be regarded as a linear combination of the $S_{n,j}(\xi)$.

<u>ORTHOGONAL TRANSFORMATIONS</u>

Suppose now that the functions $S_{n,j}(\xi)$, $j = 1,\ldots,N$ constitute an orthonormal set, i.e.,

(13) $$\int_{\Omega_q} S_{n,j}(\xi) \, S_{n,k}(\xi) \, d\omega_q = \delta_{jk} \quad .$$

If $\mathbf{A}$ is an orthogonal matrix, then $H_n(\mathbf{A}x)$ is a homogeneous harmonic polynomial of degree n in x if $H_n(x)$ has this property, so that $S_n(\mathbf{A}\xi)$ is a spherical harmonic of order n. In particular

(14) $$S_{n,j}(\mathbf{A}\xi) = \sum_{\tau=1}^{N(q,n)} c_{j\tau}^n \, S_{n,\tau}(\xi) \quad .$$

To every orthogonal matrix $\mathbf{A}$ there corresponds therefore a matrix $c_{j\tau}^n$. We now have, because of (13) and (14),

(15) $$\int_{\Omega_q} S_{n,j}(\mathbf{A}\xi) \, S_{n,k}(\mathbf{A}\xi) \, d\omega_q = \sum_{\tau=1}^{N(q,n)} c_{j\tau}^n \, c_{k\tau}^n \quad .$$

The orthogonal transformation $\mathbf{A}\xi$ may be regarded as a coordinate transformation of Ω_q which leaves the surface element $d\omega_q$ unaltered. This means that

$$\int_{\Omega_q} S_{n,j}(\mathbf{A}\xi)\, S_{n,k}(\mathbf{A}\xi)\, d\omega_q = \int_{\Omega_q} S_{n,j}(\xi)\, S_{n,k}(\xi)\, d\omega_q = \delta_{jk} .$$

From (15) we now get

$$\sum_{\tau=1}^{N(q,n)} c_{j\tau}^{n}\, c_{k\tau}^{n} = \delta_{jk} \tag{16}$$

so that the coefficients $c_{j\tau}^{n}$ are the elements of an orthogonal matrix. Besides (16) we therefore get also

$$\sum_{\tau=1}^{N(q,n)} c_{\tau j}^{n}\, c_{\tau k}^{n} = \delta_{jk} . \tag{17}$$

For any two points ξ and η on Ω_q we now form the function

$$F(\xi,\eta) = \sum_{j=1}^{N(q,n)} S_{n,j}(\xi)\, S_{n,j}(\eta) .$$

Due to (17) we have for any orthogonal matrix $\mathbf{A}$

$$\begin{aligned} F(\mathbf{A}\xi, \mathbf{A}\eta) &= \sum_{j=1}^{N(q,n)} S_{n,j}(\mathbf{A}\xi)\, S_{n,j}(\mathbf{A}\eta) \\ &= \sum_{j=1}^{N(q,n)} \Big[\sum_{\tau=1}^{N(q,n)} c_{j\tau}^{n}\, S_{n,\tau}(\xi)\Big]\cdot\Big[\sum_{m=1}^{N(q,n)} c_{jm}^{n}\, S_{n,m}(\eta)\Big] \\ &= \sum_{\tau=1}^{N(q,n)} \sum_{m=1}^{N(q,n)} \delta_{\tau m}\, S_{n,\tau}(\xi)\, S_{n,m}(\eta) = F(\xi,\eta) . \end{aligned}$$

The function $F(\xi,\eta)$ thus has the important property that it is not changed if ξ and η undergo an orthogonal transformation simultaneously.

To further studies of our function $F(\xi,\eta)$ we use the following properties of the group of orthogonal transformations

a) To every unit vector ξ there is an orthogonal transformation such that $\mathbf{A}\xi = \varepsilon_q$.

b) For any two vectors ξ and η we have
$$\xi\cdot\eta = \mathbf{A}\xi\cdot\mathbf{A}\eta .$$

c) For any unit vector ξ there is a subgroup of orthogonal transformations, which keeps ξ fixed and which transforms a given unit vector η_0 in all those vectors η for which

$$\xi \cdot \eta = \xi \cdot \eta_0 \ .$$

LEGENDRE FUNCTIONS

We now use these properties to study our function $F(\xi, \eta)$. It follows from (a) that we may transform ξ into ε_q. Then, according to (2), η would be represented in the form

$$\eta = t\,\varepsilon_q + \sqrt{1-t^2}\,\eta_{q-1} \ ; \quad t = \eta \cdot \varepsilon_q \ . \tag{18}$$

From (b) we know that t is also the value of the scalar product of ξ and η before carrying out the transformation. From (18) it can be seen that the subgroup with fixpoint ε_q is isomorphic to the orthogonal group in (q-1)-dimensions [1].

We have therefore

$$F(\varepsilon_q, t\varepsilon_q + \sqrt{1-t^2}\,\eta_{q-1}) = F(\varepsilon_q, t\varepsilon_q + \sqrt{1-t^2}\,\overset{*}{\eta}_{q-1})$$

for any two vectors η_{q-1} and $\overset{*}{\eta}_{q-1}$ on Ω_{q-1}. This implies that $F(\varepsilon_q, t\varepsilon_q + \sqrt{1-t^2}\,\eta_{q-1})$ does not depend on η_{q-1}. It therefore is a function of t alone. Combining this with (18) we have

Lemma 5: Let $S_{n,j}(\xi)$, $j = 1,\ldots,N$ be an orthonormal set of spherical harmonics on Ω_q. Then for any two points (vectors) ξ and η on Ω_q the function

$$F(\xi,\eta) = \sum_{j=1}^{N(q,n)} S_{n,j}(\xi)\, S_{n,j}(\eta) = \phi(\xi\cdot\eta)$$

depends only on the scalar product of ξ and η.

---

[1] The orthogonal group in one dimension consists of the two transformations $x'_1 = \pm x_1$ only.

It is clear from the left hand side that this function is a spherical harmonic in ξ or η of degree n. From the right hand side it follows that it is symmetric with regard to all orthogonal transformations which leave ξ fixed. We are thus led to introduce a special spherical harmonic which has this same symmetry.

<u>Definition 2</u>: Let $L_n(x)$ be a homogeneous, harmonic polynomial of degree n with the following properties:

a) $L_n(\mathbf{A}x) = L_n(x)$ for all orthogonal transformations $\mathbf{A}$ which leave the vector ε_q unchanged.

b) $L_n(\varepsilon_q) = 1$.

Then

$$\frac{1}{r^n} L_n(x) = L_n(\xi)$$

is called the Legendre function of degree n.

By this definiton the function $L_n(\xi)$ is uniquely determined, for according to the representation (4), $L_n(x)$ is uniquely determined by the homogeneous polynomials $A_n(x_1,\ldots,x_{q-1})$ and $A_{n-1}(x_1,\ldots,x_{q-1})$. The condition (a) implies that these polynomials depend only on $(x_1)^2 + (x_2)^2 + \ldots + (x_{q-1})^2$. We thus get

$$A_n = c\,[(x_1)^2 + \cdots + (x_{q-1})^2]^k;\ A_{n-1} = 0 \text{ for } n = 2k$$

and

$$A_{n-1} = c\,[(x_1)^2 + \cdots + (x_{q-1})^2]^k;\ A_n = 0 \text{ for } n = 2k+1 .$$

Apart from a multiplicative constant, the function $L_n(x)$ is therefore determined by condition (a). The value of the constant c is then fixed by condition (b). Using the parameter representation (2) we see that $L_n(\xi)$ depends on t only, as

$$(x_1)^2 + (x_2)^2 + \cdots + (x_{q-1})^2 = r^2(1-t^2) .$$

We now have:

<u>Theorem 1</u>: The Legendre function $L_n(\xi)$ may be written as

$$L_n(\xi) = P_n(t)$$

where $P_n(t)$ is a polynomial of degree n with

$$P_n(1) = 1 \; ; \; P_n(-t) = (-1)^n P_n(t) \; .$$

The last two relations of this theorem can be proved easily:

As $r = 1$, $t = 1$, corresponds to $\xi = \varepsilon_q$, the first statement is condition (b) of Definition 2 and the second equation follows from Lemma 1.

ADDITION THEOREM

We now can determine the function $\phi(\xi \cdot \eta)$ in Lemma 5, for we know that this function is a spherical harmonic of degree n with respect to η. It is moreover unchanged if η is transformed by an orthogonal transformation which leaves ξ fixed, so that

$$\sum_{j=1}^{N(q,n)} S_{n,j}(\xi)\, S_{n,j}(\eta) = c_n P_n(\xi \cdot \eta) \; ,$$

as the function $\phi(\xi \cdot \eta)$ can only be proportional to $P_n(\xi \cdot \eta)$. To determine the constant c_n we set $\xi = \eta$ and obtain

$$\sum_{j=1}^{N(q,n)} [S_{n,j}(\xi)]^2 = c_n P_n(1) = c_n \; .$$

Integration over Ω_q gives

$$N(q,n) = c_n \omega_q$$

and we get

Theorem 2 : (Addition Theorem) Let $S_{n,j}(\xi)$ be an orthonormal set of N(q,n) spherical harmonics of order n and dimension q. Then

$$\sum_{j=1}^{N(q,n)} S_{n,j}(\xi)\, S_{n,j}(\eta) = \frac{N(q,n)}{\omega_q} P_n(\xi\cdot\eta) ,$$

where $P_n(t)$ is the Legendre Polynomial of degree n and dimension q.

This theorem is called addition theorem as it reduces to the addition theorem for the function cos φ in the two-dimensional case after introducing polar coordinates.

In order to determine the spherical harmonics for the case q = 2 according to this theory we first have to determine two linearly independent homogeneous and harmonic polynomials of degree n. We can take them as

$$Re\,(x_2 + i x_1)^n \qquad \text{and} \qquad Im\,(x_2 + i x_1)^n .$$

We now introduce a system of polar coordinates in the usual way

(19) $$x_1 = r\cos\varphi \quad ; \quad x_2 = r\sin\varphi$$

and get

$$\frac{1}{r^n} Re\,(x_2 + i x_1)^n = \cos n\left(\frac{\pi}{2} - \varphi\right)$$

$$\frac{1}{r^n} Im\,(x_2 + i x_1)^n = \sin n\left(\frac{\pi}{2} - \varphi\right) .$$

From these two we get an orthonormal set by

$$S_{n,1} = \frac{1}{\sqrt{\pi}} \cos n\left(\frac{\pi}{2} - \varphi\right) , \quad S_{n,2} = \frac{1}{\sqrt{\pi}} \sin n\left(\frac{\pi}{2} - \varphi\right) .$$

The Legendre function now is obtained from a homogeneous harmonic polynomial which is symmetric with respect to the x_2- axis, and which takes on the value 1 for $x_1 = 0$, $x_2 = 1$. This gives us

$$L_n(x_1, x_2) = Re\,(x_2 + i x_1)^n$$

or

$$L_n(\xi) = \cos n\left(\frac{\pi}{2} - \varphi\right) .$$

Now let t be the scalar product between ε_2 and ξ. We then have from (19)

$$t = \sin\varphi = \cos\left(\frac{\pi}{2} - \varphi\right)$$

which gives us

$$L_n(\xi) = \cos n\,(\cos^{-1} t) = P_n(t) .$$

In two dimensions, therefore, the function $P_n(t)$ is what is otherwise known as the Chebychev Polynomial.

If the points ξ and η have the coordinates φ and ψ respectively we get by observing that

$$\xi \cdot \eta = \cos(\varphi - \psi) ; \quad N(2,n) = 2 , \quad n \geq 1 ; \quad \omega_2 = 2\pi$$

the relation (for q = 2):

$$\sum_{j=1}^{2} S_{n,j}(\xi)\, S_{n,j}(\eta) = \frac{1}{\pi}\left[\cos n\left(\frac{\pi}{2}-\varphi\right)\cdot \cos n\left(\frac{\pi}{2}-\psi\right) + \sin n\left(\frac{\pi}{2}-\varphi\right)\cdot \sin n\left(\frac{\pi}{2}-\psi\right)\right]$$

$$= \frac{1}{\pi}\cos n(\varphi - \psi) = \frac{1}{\pi} P_n(\cos(\varphi - \psi)) .$$

Theorem 2 therefore reduces to the addition formula for the function $\cos\varphi$ in the two-dimensional case, which explains why this result is called the addition theorem of spherical harmonics.

REPRESENTATION THEOREM

As is well known, all the trigonometric functions can be derived by simple algebraic processes from a single one (e.g.cosx), the question arises if there is a corresponding result in the theory of general spherical harmonics. The addition theorem suggests that it might be possible to express all spherical harmonics in terms of the Legendre function. This is stated in

<u>Theorem 3</u>: To every degree n, there is a system of N points $\eta_1, \eta_2, \ldots\ldots, \eta_N$ such that every spherical harmonic $S_n(\xi)$ can be expressed in the form

$$S_n(\xi) = \sum_{k=1}^{N(q,n)} a_k P_n(\eta_k \cdot \xi) .$$

It is clear from the above that every spherical harmonic can be written as

$$S_n(\xi) = \sum_{j=1}^{N(q,n)} c_j S_{n,j}(\xi)$$

so that it is only necessary to show that the functions $S_{n,j}(\xi)$ can be expressed by the Legendre functions.

To this end we observe that it is certainly possible to find a point η_1 such that $S_{n,1}(\eta_1) \neq 0$. We then consider

$$\begin{vmatrix} S_{n,1}(\eta_1) & S_{n,1}(\xi) \\ S_{n,2}(\eta_1) & S_{n,2}(\xi) \end{vmatrix} .$$

As a function of ξ this cannot be identically 0, because $S_{n,1}(\xi)$ and $S_{n,2}(\xi)$ are linearly independent. Therefore there is a point $\xi = \eta_2$ such that this determinant does not vanish. Discussing next the determinant

$$\begin{vmatrix} S_{n,1}(\eta_1) & S_{n,1}(\eta_2) & S_{n,1}(\xi) \\ S_{n,2}(\eta_1) & S_{n,2}(\eta_2) & S_{n,2}(\xi) \\ S_{n,3}(\eta_1) & S_{n,3}(\eta_2) & S_{n,3}(\xi) \end{vmatrix}$$

and using the same arguments we obtain by induction

<u>Lemma 6</u> : There is a system of points $\eta_1, \eta_2, \cdots, \eta_N$ such that the matrix $(S_{n,j}(\eta_k))$, $j = 1,\ldots,N$; $k = 1,\ldots,N$ is non-degenerate.

From Theorem 2 we now have

$$\sum_{j=1}^{N(q,n)} S_{n,j}(\eta_k)\, S_{n,j}(\xi) = \frac{N(q,n)}{\omega_q} P_n(\eta_k \cdot \xi)$$

This is a non-degenerate system of linear equations with $S_{n,j}(\xi)$ as unknowns so that Theorem 3 follows by inversion.

In order to simplify the formulation fo these relations we introduce

<u>Definition 3</u> : A system of N points $\eta_1, \cdots, \eta_N$ on Ω_q will be called a fundamental system of degree n, if

$$\det \{ P_n(\eta_i \cdot \eta_k) \} > 0 .$$

It can be seen readily that the matrix $\frac{N}{\omega_q} P_n(\eta_i \cdot \eta_k)$ can be obtained by multiplying the matrix $S_{n,j}(\eta_i)$ with its adjoint, so that the determinant of Definition 3 is non-negative. If the determinant is positive, the system has the properties stated in Theorem 3, since then det $(S_{n,j}(\eta_k)) \neq 0$, which may also be formulated as

<u>Theorem 4</u> : Every spherical harmonic of degree n may be represented in the form

$$S_n(\xi) = \sum_{k=1}^{N(q,n)} a_k P_n(\eta_k \cdot \xi)$$

if the points η_k form a fundamental system of this degree.

It is clear now that an orthonormal system of spherical harmonics can always be obtained by linear combinations of the functions $P_n(\eta_k \cdot \xi)$. Which fundamental system η_k is best suited to represent the functions of degree n remains open at this stage as it requires more information on the polynomials $P_n(t)$.

APPLICATIONS OF THE ADDITION THEOREM

Before studying the Legendre polynomials in detail, we shall obtain several simple results on spherical harmonics in general which depend on the addition theorem.

If we remember that every spherical harmonic of degree n can be represented as

$$(20)\qquad S_n(\xi) = \sum_{k=1}^{N(q,n)} a_k\, S_{n,k}(\xi) \quad , \quad a_k = \int_{\Omega_q} S_n(\eta)\, S_{n,k}(\eta)\, d\omega_q$$

we get immediately from Theorem 2

Lemma 7 : For every spherical harmonic of degree n

$$\frac{N(q,n)}{\omega_q} \int_{\Omega_q} P_n(\xi\cdot\eta)\, S_n(\eta)\, d\omega_q(\eta) = S_n(\xi)$$

Here the letter η in connection with $d\omega_q$ means that the integration is carried out with respect to η .

Observing that

$$\int_{\Omega_q} S_n^2(\xi)\, d\omega_q = \sum_{k=1}^{N(q,n)} (a_k)^2$$

we get from (20), using Schwarz's inequality and Theorem 2,

$$(21)\qquad |S_n(\xi)|^2 \leq \sum_{k=1}^{N(q,n)} (a_k)^2 \sum_{k=1}^{N(q,n)} [S_{n,k}(\xi)]^2 = \frac{N(q,n)}{\omega_q} P_n(1)\cdot \sum_{k=1}^{N(q,n)} (a_k)^2 .$$

This gives us

Lemma 8 : Let $S_n(\xi)$ be a spherical harmonic of degree n. Then

$$|S_n(\xi)| \leq \sqrt{\frac{N(q,n)}{\omega_q} \int_{\Omega_q} |S_n(\xi)|^2\, d\omega_q} .$$

Put

$$S_n(\xi) = \frac{N(q,n)}{\omega_q} P_n(\xi\cdot\eta) = \sum_{j=1}^{N(q,n)} S_{n,j}(\xi)\, S_{n,j}(\eta) ,$$

then we get from (21) and Theorem 2

$$\left[\frac{N(q,n)}{\omega_q}\right]^2\cdot[P_n(\xi\cdot\eta)]^2 \leq \frac{N(q,n)}{\omega_q} \sum_{k=1}^{N(q,n)} [S_{n,k}(\eta)]^2 = \left[\frac{N(q,n)}{\omega_q}\right]^2$$

which gives us

Lemma 9 : For $-1 \leq t \leq 1$ $\quad |P_n(t)| \leq 1$.

From Theorem 2 we have moreover

$$\left[\frac{N(q,n)}{\omega_q}\right]^2 \cdot \left[P_n(\xi\cdot\eta)\right]^2 = \left[\sum_{j=1}^{N(q,n)} S_{n,j}(\xi)\cdot S_{n,j}(\eta)\right]^2 .$$

This gives by integration over Ω_q

$$(22) \qquad \left[\frac{N(q,n)}{\omega_q}\right]^2 \int_{\Omega_q} \left[P_n(\xi\cdot\eta)\right]^2 d\omega_q(\eta) = \sum_{j=1}^{N(q,n)} \left[S_{n,j}(\xi)\right]^2 = \frac{N(q,n)}{\omega_q} .$$

As the value of the integral on the left hand side does not depend on ξ, we may assume ξ to be ε_q. Then, using the coordinate representation (2), we get

$$(23) \qquad \begin{aligned} \int_{\Omega_q} \left[P_n(\xi\cdot\eta)\right]^2 d\omega_q(\eta) &= \int_{\Omega_{q-1}} \int_{-1}^{+1} P_n^2(t)(1-t^2)^{\frac{q-3}{2}} dt\, d\omega_{q-1} \\ &= \omega_{q-1} \int_{-1}^{+1} P_n^2(t)(1-t^2)^{\frac{q-3}{2}} dt . \end{aligned}$$

It follows from (22) and (23)

$$(24) \qquad \int_{-1}^{+1} P_n^2(t)(1-t^2)^{\frac{q-3}{2}} dt = \frac{\omega_q}{\omega_{q-1}} \cdot \frac{1}{N(q,n)} = \frac{\sqrt{\pi}\,\Gamma\left(\frac{q-1}{2}\right)}{\Gamma\left(\frac{q}{2}\right)} \cdot \frac{1}{N(q,n)} .$$

On the other hand , by Lemma 2,

$$\int_{\Omega_q} P_n(\xi\cdot\eta)\, P_m(\xi\cdot\eta)\, d\omega_q(\eta) = 0 \qquad \text{for } n \neq m.$$

By the coordinate representation (2) this is equivalent to

$$\int_{-1}^{+1} P_n(t)\, P_m(t)\, (1-t^2)^{\frac{q-3}{2}} dt = 0 \qquad \text{for } n \neq m,$$

which gives us, combined with (24)

Lemma 10 :

$$\int_{-1}^{+1} P_n(t)\, P_m(t)\, (1-t^2)^{\frac{q-3}{2}} dt = \frac{\omega_q}{\omega_{q-1}} \cdot \frac{1}{N(q,n)} \cdot \delta_{nm} .$$

RODRIGUES' FORMULA

We shall now give a representation of the Legendre polynomials based on the following properties:

1. $P_n(t)$ is a polynomial of degree n in t.

2. $\int_{-1}^{+1} P_n(t)\, P_m(t)\, (1-t^2)^{\frac{q-3}{2}}\, dt = 0$ for $n \neq m$.

3. $P_n(1) = 1$.

The usual process of orthogonalization shows that $P_n(t)$ is determined up to a multiplicative constant by the first two conditions. This constant can then be fixed by the third condition.

Consider the functions

$$(25) \qquad f_n(q,t) = (1-t^2)^{\frac{3-q}{2}} \cdot \left(\frac{d}{dt}\right)^n (1-t^2)^{\frac{n+(q-3)}{2}} .$$

They are polynomials of degree n, and we see by partial integration that

$$\int_{-1}^{+1} f_n(q,t)\, f_m(q,t)\, (1-t^2)^{\frac{q-3}{2}}\, dt$$

$$= (-1)^n \int_{-1}^{+1} (1-t^2)^{\frac{n+(q-3)}{2}} \left(\frac{d}{dt}\right)^n f_m(q,t)\, dt , \quad (q \geq 2) .$$

If $n > m$ the right hand side vanishes, which proves that the functions (25) satisfy the first two conditions.

Put $t = 1 - s$, then

$$f_n(q,1-s) = (-1)^n\, [(2-s)\cdot s]^{\frac{3-q}{2}} \left(\frac{d}{ds}\right)^n [(2-s)\cdot s]^{\frac{n+(q-3)}{2}}$$

so that we get

$$f_n(q,1) = (-2)^n \left(n+\frac{q-3}{2}\right)\left(n+\frac{q-3}{2}-1\right) \cdots\cdots \left(\frac{q-1}{2}\right)$$

$$= (-2)^n \frac{\Gamma\left(n+\frac{q-1}{2}\right)}{\Gamma\left(\frac{q-1}{2}\right)} .$$

Thus we get

Theorem 5 : (Rodrigues' formula)

$$P_n(t) = \left(-\frac{1}{2}\right)^n \frac{\Gamma(\frac{q-1}{2})}{\Gamma(n+\frac{q-1}{2})} (1-t^2)^{\frac{3-q}{2}} \left(\frac{d}{dt}\right)^n (1-t^2)^{\frac{n+(q-3)}{2}} .$$

This has an immediate and simple application which we obtain after integrating n times by parts. It is

Lemma 11 : Let f(t) be n times continuously differentiable, then

$$\int_{-1}^{+1} f(t)\, P_n(t)\, (1-t^2)^{\frac{q-3}{2}}\, dt$$

$$= \left(\frac{1}{2}\right)^n \frac{\Gamma(\frac{q-1}{2})}{\Gamma(n+\frac{q-1}{2})} \int_{-1}^{+1} (1-t^2)^{\frac{n+(q-3)}{2}} f^{(n)}(t)\, dt .$$

As an immediate application of Lemma 11 we determine the leading coefficient of the Legendre polynomial of order n. If c_n is the coefficient of the highest power in $P_n(t)$, then

$$(26) \qquad \int_{-1}^{+1} P_n^2(t)\,(1-t^2)^{\frac{q-3}{2}}\, dt = c_n \int_{-1}^{+1} P_n(t)\, t^n (1-t^2)^{\frac{q-3}{2}}\, dt ,$$

as the lower terms of the power series for $P_n(t)$ do not contribute to the integral. The left hand side of (26) is $\frac{\omega_q}{\omega_{q-1}} \cdot \frac{1}{N(q,n)}$ according to (24) and the right hand side equals

$$c_n \left(\tfrac{1}{2}\right)^n \frac{\Gamma\left(\frac{q-1}{2}\right)}{\Gamma\left(n+\frac{q-1}{2}\right)} \; n! \int_{-1}^{+1} (1-t^2)^{\frac{n+q-3}{2}} dt$$

$$= c_n \left(\tfrac{1}{2}\right)^n \frac{\Gamma\left(\frac{q-1}{2}\right)}{\Gamma\left(n+\frac{q-1}{2}\right)} \; n! \int_{0}^{1} (1-t^2)^{\frac{n+(q-3)}{2}} \, t^{-\frac{1}{2}} dt$$

$$= c_n \, n! \left(\tfrac{1}{2}\right)^n \frac{\Gamma\left(\frac{q-1}{2}\right)}{\Gamma\left(n+\frac{q-1}{2}\right)} \quad \frac{\Gamma\left(n+\frac{q-1}{2}\right) \cdot \Gamma\left(\frac{1}{2}\right)}{\Gamma\left(n+\frac{q}{2}\right)} .$$

Therefore

$$P_n(t) = \frac{\omega_q}{\omega_{q-1}} \cdot \frac{1}{\sqrt{\pi}\, N(q,n)} \cdot \frac{\Gamma\left(n+\frac{q}{2}\right)}{\Gamma\left(\frac{q-1}{2}\right)} \; \frac{2^n}{n!} \, t^n + \cdots .$$

By (3)

$$\frac{\omega_q}{\omega_{q-1}} = \frac{\sqrt{\pi}\, \Gamma\left(\frac{q-1}{2}\right)}{\Gamma\left(\frac{q}{2}\right)}$$

so that

$$P_n(t) = \frac{1}{N(q,n)} \cdot \frac{\Gamma\left(n+\frac{q}{2}\right)}{\Gamma\left(\frac{q}{2}\right)} \; \frac{2^n}{n!} \, t^n + \cdots . \tag{27}$$

FUNK - HECKE FORMULA

Before going further into the details of the Legendre functions we shall discuss a formula which will prove to be the basis of a great many special results.

Let us consider an integral of the form

$$F(\alpha,\beta) = \int_{\Omega_q} f(\alpha\cdot\eta) \, P_n(\beta\cdot\eta) \, d\omega_q(\eta)$$

where $f(t)$ is a continuous function for $-1 \leq t \leq 1$ and the integration is carried out with respect to η. Then with any orthogonal matrix **A**

(28)
$$F(\mathbf{A}\alpha, \mathbf{A}\beta) = \int_{\Omega_q} f(\mathbf{A}\alpha\cdot\eta)\, P_n(\mathbf{A}\beta\cdot\eta)\, d\omega_q(\eta)$$

$$= \int_{\Omega_q} f(\alpha\cdot\mathbf{A}^*\eta)\, P_n(\beta\cdot\mathbf{A}^*\eta)\, d\omega_q(\eta)$$

where $\mathbf{A}^*$ is the adjoint (transpose) of $\mathbf{A}$. Now the surface elements $d\omega_q(\mathbf{A}^*\eta)$ and $d\omega_q(\eta)$ are equal so that (28) becomes

$$F(\mathbf{A}\alpha, \mathbf{A}\beta) = \int_{\Omega_q} f(\alpha\cdot\mathbf{A}^*\eta)\, P_n(\beta\cdot\mathbf{A}^*\eta)\, d\omega_q(\mathbf{A}^*\eta) .$$

This is equal to $F(\alpha,\beta)$ because we may regard $\mathbf{A}^*\cdot\eta$ as the new variables. Using the same argument now which led to Lemma 5, we see that $F(\alpha,\beta)$ is a function of the scalar product only, which gives us

$$\int_{\Omega_q} f(\alpha\cdot\eta)\, P_n(\beta\cdot\eta)\, d\omega_q(\eta) = \Phi(\alpha\cdot\beta) .$$

Now as a function of β this is a spherical harmonic of degree n. As it depends on the scalar product only, it has the same symmetry which characterizes $P_n(\alpha\cdot\beta)$. Therefore we get

$$\int_{\Omega_q} f(\alpha\cdot\eta)\, P_n(\beta\cdot\eta)\, d\omega_q(\eta) = \lambda\, P_n(\alpha\cdot\beta) .$$

In order to determine λ set $\alpha = \beta = \epsilon_q$ and

$$\eta = \eta_q = t\,\epsilon_q + \sqrt{1-t^2}\;\eta_{q-1} .$$

Then with

$$d\omega_q = (1-t^2)^{\frac{q-3}{2}}\, d\omega_{q-1}$$

we get

$$\lambda = \lambda\, P_n(1) = \int_{\Omega_{q-1}} \int_{-1}^{+1} f(t)\, P_n(t)\, (1-t^2)^{\frac{q-3}{2}}\, dt\, d\omega_{q-1}$$

$$= \omega_{q-1} \int_{-1}^{+1} f(t)\, P_n(t)\, (1-t^2)^{\frac{q-3}{2}}\, dt .$$

This leads to

Lemma 12 : Let α and β be any two points in Ω_q , and suppose f(t) is continuous for $-1 \leq t \leq 1$. Then

$$\int_{\Omega_q} f(\alpha\cdot\eta)\, P_n(\beta\cdot\eta)\, d\omega_q(\eta) = \lambda\, P_n(\alpha\cdot\beta)$$

where

$$\lambda = \omega_{q-1} \int_{-1}^{+1} f(t)\, P_n(t)\, (1-t^2)^{\frac{q-3}{2}}\, dt .$$

From Lemma 7 we now get by multiplication with $S_n(\beta)$ and integration with regard to β

Theorem 6 : (Funk-Hecke formula) Suppose f(t) is continuous for $-1 \leq t \leq 1$. Then for every spherical harmonic of degree n

$$\int_{\Omega_q} f(\alpha\cdot\eta)\, S_n(\eta)\, d\omega_q(\eta) = \lambda\, S_n(\alpha)$$

with

$$\lambda = \omega_{q-1} \int_{-1}^{+1} f(t)\, P_n(t)\, (1-t^2)^{\frac{q-3}{2}}\, dt .$$

INTEGRAL REPRESENTATIONS OF SPHERICAL HARMONICS

To distinguish clearly we will designate in the following a spherical harmonic of order n in q dimensions with $S_n(q;\xi)$ and the Legendre polynomial of degree n in q dimensions with $P_n(q;t)$.

It is obvious that the integral

$$\int_{\Omega_{q-1}} (x\cdot\varepsilon_q + i x\cdot\eta_{q-1})^n \cdot f(\eta_{q-1})\, d\omega_{q-1}(\eta)$$

represents a homogeneous harmonic polynomial of degree n for any continous function $f(\eta_{q-1})$, if we set

$$\eta = \eta_q = u\,\varepsilon_q + \sqrt{1-u^2}\;\eta_{q-1}$$

where

$$u = \eta\cdot\varepsilon_q$$

This enables us to get a new representation of the Legendre polynomials. To this end we now prove the identity

$$\frac{1}{\omega_{q-1}}\int_{\Omega_{q-1}} (x\cdot\varepsilon_q + i x\cdot\eta_{q-1})^n\, d\omega_{q-1}(\eta) = L_n(q,x)\,.$$

As this integral represents the average over all directions which are perpendicular to ε_q, the integral is symmetric with respect to all orthogonal transformations which leave ε_q fixed. For $x = \varepsilon_q$ the integral assumes the value one; hence the integral satisfies Definition 2. We therefore have

$$L_n(q,\xi) = P_n(q,t) = \frac{1}{\omega_{q-1}}\int_{\Omega_{q-1}} (\xi\cdot\varepsilon_q + i\xi\cdot\eta_{q-1})^n\, d\omega_{q-1}(\eta)$$

Now

$$\xi = \xi_q = t\,\varepsilon_q + \sqrt{1-t^2}\;\xi_{q-1}$$

so that

$$P_n(q,t) = \frac{1}{\omega_{q-1}}\int_{\Omega_{q-1}} (t + i\sqrt{1-t^2}\,[\xi_{q-1}\cdot\eta_{q-1}])^n\, d\omega_{q-1}(\eta)$$

and Theorem 6 with $S_0 = 1$ gives

Theorem 7 : (Laplace's representation)

$$P_n(q,t) = \frac{\omega_{q-2}}{\omega_{q-1}} \int_{-1}^{+1} (t + i\sqrt{1-t^2}\cdot s)^n \cdot (1-s^2)^{\frac{q-4}{2}}\, ds .$$

Similarly we may get representations for further spherical harmonic functions if we consider

$$\int_{\Omega_{q-1}} (x\cdot\varepsilon_q + i x\cdot\eta_{q-1})^n \, S_j(q-1,\eta_{q-1})\, d\omega_{q-1}(\eta) .$$

For x = ξ this becomes a spherical harmonic of degree n in q dimensions which we may represent in the form

$$\int_{\Omega_{q-1}} (t + i\sqrt{1-t^2}\,[\xi_{q-1}\cdot\eta_{q-1}])^n \, S_j(q-1,\eta_{q-1})\, d\omega_{q-1}(\eta) .$$

According to Hecke's formula (Theorem 6) this is

(29) $$S_j(q-1,\xi_{q-1})\, \omega_{q-2} \int_{-1}^{+1} (t + i\sqrt{1-t^2}\cdot s)^n \, P_j(q-1,s)\,(1-s^2)^{\frac{q-4}{2}}\, ds$$

which can be written as

$$A_{n,j}(q,t)\; S_j(q-1,\xi_{q-1}) .$$

ASSOCIATED LEGENDRE FUNCTIONS

In order to get an explicit representation of a system of orthonormal spherical harmonics we now introduce

Definition 4 : Suppose the points of Ω_q are represented in the form

$$\xi_q = t\cdot\varepsilon_q + \sqrt{1-t^2}\; \xi_{q-1} .$$

Then, the function $A_{n,j}(q,t)$ is called an associated Legendre function of degree n, order j, and dimension q, if

$$A_{n,j}(q,t)\,S_j(q-1;\xi_{q-1})\quad,\quad j=0,1,\dots,n$$

is a spherical harmonic of degree n in q dimensions for every spherical harmonic $S_j(q-1, \xi_{q-1})$ of degree j in q-1 dimensions.

The functions $A_{n,j}(q,t)$ will be called normalized if

$$\int_{-1}^{+1} A_{n,j}(q,t)\,A_{m,j}(q,t)\,(1-t^2)^{\frac{q-3}{2}}\,dt = \delta_{nm}$$

As sperical harmonics of different degree are orthogonal we only have to determine the factor of normalization for the case n = m.

The associated Legendre functions of order zero are readily obtained, because Definition 4 means in this case that the corresponding special harmonics have the symmetry properties which determined $P_n(q,t)$. Therefore $A_{n,0}(q,t)$ and $P_n(q,t)$ are proportional, and we have to determine the constant. This gives

$$(30)\qquad A_{n,0}(q,t) = \sqrt{\frac{N(q,n)\,\omega_{q-1}}{\omega_q}}\;P_n(q,t)\,.$$

From (29) it is obvious that

$$\int_{-1}^{+1}\left(t+i\sqrt{1-t^2}\cdot s\right)^n P_j(q-1,s)(1-s^2)^{\frac{q-4}{2}}\,ds$$

is an associated Legendre function of degree n, order j and dimension q. From Lemma 11 we now see that apart from a multiplocative constant this is equal to

$$(1-t^2)^{j/2}\int_{-1}^{+1}\left(t+i\sqrt{1-t^2}\cdot s\right)^{n-j}(1-s^2)^{\frac{j+q-4}{2}}\,ds.$$

The integral is proportional to $P_{n-j}(2j + q,t)$ as follows immediately from Theorem 7, so that

$$(1-t^2)^{j/2}\;P_{n-j}(2j+q,t)$$

is an associated Legendre function of degree n and order j in q dimensions.

Consider now the function

$$\left(\frac{d}{dt}\right)^j P_n(q,t) = P_n^{(j)}(q,t)$$

which is a polynomial of degree n-j. Integrating j times by parts, we see that for $n > m$, $j = 0,\dots, m$; $q \geq 3$

$$\int_{-1}^{+1} (1-t^2)^j P_n^{(j)}(q,t)\, P_m^{(j)}(q,t)\, (1-t^2)^{\frac{q-3}{2}}\, dt \tag{31}$$

$$= (-1)^j \int_{-1}^{+1} P_n(q,t) \left(\frac{d}{dt}\right)^j \left[(1-t^2)^{\frac{j+(q-3)}{2}} P_m^{(j)}(q,t)\right] dt$$

because the integrated terms vanish for t = 1 and t = -1. The differentiated term is of the form

$$(1-t^2)^{\frac{q-3}{2}}\; p_m(t) \tag{32}$$

where $p_m(t)$ is a polynomial of degree m, which is best seen by using the formula

$$\left(\frac{d}{dt}\right)^j U\cdot V = \sum_{k=0}^{j} \binom{j}{k} U^{(k)} V^{(j-k)}$$

with $U = (1-t^2)^{j+(q-3)/2}$ and $V = P_m^{(j)}(q,t)$.

We thus obtain from (31) for $m \neq n$

$$\int_{-1}^{+1} (1-t^2)^{\frac{q-3+2j}{2}}\; P_n^{(j)}(q,t)\, P_m^{(j)}(q,t)\, dt = 0 . \tag{33}$$

On the other hand we have from Lemma 10 for $m \neq n$

$$\int_{-1}^{+1} (1-t^2)^{\frac{q-3+2j}{2}}\; P_{n-j}(q+2j,t)\, P_{m-j}(q+2j,t)\, dt = 0 . \tag{34}$$

As $P_n^{(j)}(q,t)$ and $P_{n-j}(q+2j,t)$ are both polynomials of degree n-j which satisfy the same conditions of orthogonality, they may be obtained by a process of orthogonalization from the powers t^n with the weight function $(1-t^2)^{\frac{q-3+2j}{2}}$ over the intervall $-1\leq t\leq 1$. As they are not normalized they differ only by a constant factor. This gives us

Lemma 13 : The functions

$$(1-t^2)^{j/2} \; P_{n-j}(2j+q,t)$$

and

$$(1-t^2)^{j/2} \; P_n^{(j)}(q,t)$$

are associated Legendre functions of degree n, order j, and dimension q, which differ only by a factor of normalization, which is given by

$$P_{n-j}(2j+q,t) = \frac{N(q,n)}{N(2j+q,n-j)} \cdot \frac{\Gamma(\frac{q}{2})}{\Gamma(j+\frac{q}{2})} \cdot 2^{-j} \, P_n^{(j)}(q,t) .$$

As $P_{n-j}(2j+q,t)$ and $P_n^{(j)}(q,t)$ are proportional, this last result can be obtained by equating the coefficients of t^{n-j} as given by (27). This shows also that all Legendre polynomials can be expressed either by $P_n(3,t)$ or $P_n(2,t)$ according to whether q is odd or even.

The purpose of the preceding study, however, was not to find a relation between Legendre polynomials of different dimensions but to give an explicit representation of the normalized associated Legendre functions $A_{n,j}(q,t)$.

Suppose now that the two unit vectors ξ and η are represented in the form

$$\xi = t \cdot \varepsilon_q + \sqrt{1-t^2}\; \xi_{q-1}$$

$$\eta = s \cdot \varepsilon_q + \sqrt{1-s^2}\; \eta_{q-1} .$$

Then, if the function $A_{n,j}$ are normalized

$$A_{n,j}(q,t) \; S_{j,k}(q-1,\xi_{q-1}) \quad ; \quad j=0,\dots,n; \; k=1,\dots,N(q-1,j)$$

is a complete and normalized system of spherical harmonics of order n, as $S_{j,k}(q-1,\xi_{q-1})$ has this property in (q-1) dimensions because of

$$N(q,n) = \sum_{j=0}^{n} N(q-1, j) .$$

Thus we know that $A_{n,j}(q,t)$ is proportional to $(1-t^2)^{j/2}P_{n-j}(2j+q,t)$ or by Lemma (13) to $(1-t^2)^{j/2}P_n^{(j)}(q,t)$. Now

(35)
$$\int_{-1}^{+1} [(1-t^2)^{j/2} P_{n-j}(2j+q,t)]^2 (1-t^2)^{\frac{q-3}{2}} dt$$
$$= \int_{-1}^{+1} P^2_{n-j}(2j+q,t)(1-t^2)^{\frac{2j+q-3}{2}} dt$$

$$= \frac{\omega_{2j+q}}{\omega_{2j+q-1}} \cdot \frac{1}{N(2j+q, n-j)} = \frac{\sqrt{\pi}\,\Gamma(j+\frac{q-1}{2})}{\Gamma(j+\frac{q}{2})} \cdot \frac{1}{N(2j+q, n-j)} .$$

To find the normalizing factor for $(1-t^2)^{j/2}P_n^{(j)}(q,t)$, we observe that, for large t, $P_n^{(j)}(q,t)(1-t^2)^{j+(q-3)/2}$ is a holomorphic function of t which may be written as

$$P_n^{(j)}(q,t)\,(1-t^2)^{\frac{j+(q-3)}{2}} = (-1)^{j/2+\frac{q-3}{2}}\, t^{2j+q-3}\, P_n^{(j)}(q,t)\,(1-t^{-2})^{\frac{j+(q-3)}{2}} .$$

According to (27), the highest power of the Laurent expansion for $|t| > 1$ is

$$(-1)^{\frac{j+(q-3)}{2}} \frac{b_n}{(n-j)!}\, t^{n+j+q-3}$$

where we have set $b_n/n!$ for the leading coefficient in (27). Thus by (32)

$$\left(\frac{d}{dt}\right)^j \left[P_n^{(j)}(q,t)\,(1-t^2)^{\frac{j+(q-3)}{2}} \right] = (1-t^2)^{\frac{q-3}{2}} P_n(t)$$

$$= (-1)^{\frac{j+(q-3)}{2}} \frac{b_n}{(n-j)!} \frac{\Gamma(n+j+q-2)}{\Gamma(n+q-2)}\, t^{n+q-3} + \cdots .$$

Now with a constant c we get for $|t| > 1$

$$p_n(t) = (1-t^2)^{\frac{3-q}{2}} \left[(-1)^{\frac{j+(q-3)}{2}} c \cdot t^{n+q-3} + \dots \right]$$

$$= (1-t^{-2})^{\frac{3-q}{2}} \left[(-1)^j \cdot c \cdot t^n + \dots \right]$$

$$= (-1)^j \frac{b_n}{(n-j)!} \frac{\Gamma(n+j+q-2)}{\Gamma(n+q-2)} t^n + \dots$$

As $p_n(t)$ is a polynomial of degree n, we have

$$\left(\frac{d}{dt}\right)^j \left[P_n^{(j)}(q,t)(1-t^2)^{\frac{j+(q-3)}{2}} \right] =$$

$$= (1-t^2)^{\frac{q-3}{2}} \left[(-1)^j \frac{b_n}{(n-j)!} \cdot \frac{\Gamma(n+j+q-2)}{\Gamma(n+q-2)} t^n + \dots \right]$$

Substituting this into (31), we have for the value of that integral

$$\frac{b_n}{(n-j)!} \cdot \frac{\Gamma(n+j+q-2)}{\Gamma(n+q-2)} \int_{-1}^{+1} (1-t^2)^{\frac{q-3}{2}} P_n(q,t)\, t^n\, dt .$$

From the analysis leading to formula (27) we know that this last integral is

$$\sqrt{\pi}\; n!\; 2^{-n} \frac{\Gamma(\frac{q-1}{2})}{\Gamma(n+\frac{q}{2})}$$

which combined with the value of b_n from (27) yields

<u>Lemma 14</u> :

$$\int_{-1}^{+1} \left[(1-t^2)^{j/2} P_n^{(j)}(q,t) \right]^2 (1-t^2)^{\frac{q-3}{2}} dt$$

$$= \frac{\omega_q}{\omega_{q-1}} \cdot \frac{n!}{(n-j)!} \cdot \frac{\Gamma(n+j+q-2)}{\Gamma(n+q-2)} \cdot \frac{1}{N(q,n)}$$

Thus from (35) and Lemma 14 we get

<u>Lemma 15</u> : The functions

$$A_{n,j}(q,t) = \sqrt{\frac{N(2j+q,n-j)\,\omega_{2j+q-1}}{\omega_{2j+q}}}\,(1-t^2)^{j/2} P_{n-j}(2j+q,t)$$

or

$$A_{n,j}(q,t) = \sqrt{\frac{\omega_{q-1}}{\omega_q} \frac{(n-j)!}{n!} \frac{\Gamma(n+q-2)}{\Gamma(n+j+q-2)} N(q,n)} \, (1-t^2)^{j/2} P_n^{(j)}(q,t)$$

form a system of normalized associated Legendre functions.

A representation of $A_{n,j}(q,t)$ in terms of $P_n^{(j)}(q,t)$ could have been obtained from Lemma 13 but Lemma 14 is an interesting formula itself. The reader may find it interesting to compare the coefficient above with that obtained by using Lemma 13.

The addition theorem (Theorem 2) now can be written in the form

$$\sum_{j=0}^{n} A_{n,j}(q,t) \, A_{n,j}(q,s) \sum_{k=1}^{N(q-1,j)} S_{j,k}(q-1,\xi_{q-1}) \, S_{j,k}(q-1,\eta_{q-1})$$

$$= \frac{N(q,n)}{\omega_q} P_n(q, ts + \sqrt{1-t^2}\sqrt{1-s^2}\,[\xi_{q-1}\cdot\eta_{q-1}]) \, .$$

According to Theorem 2,

$$(36) \qquad \sum_{k=1}^{N(q-1,j)} S_{j,k}(q-1,\xi_{q-1}) \, S_{j,k}(q-1,\eta_{q-1}) = \frac{N(q-1,j)}{\omega_{q-1}} P_j(q-1, \xi_{q-1}\cdot\eta_{q-1}) \, ,$$

this may be written as

$$(37) \qquad \frac{1}{\omega_{q-1}} \sum_{j=0}^{n} A_{n,j}(q,t) \, A_{n,j}(q,s) \, N(q-1,j) \, P_j(q-1, \xi_{q-1}\cdot\eta_{q-1})$$
$$= \frac{N(q,n)}{\omega_q} P_n(q, t\cdot s + \sqrt{1-t^2}\sqrt{1-s^2}\,[\xi_{q-1}\cdot\eta_{q-1}]) \, .$$

The addition theorem is usually given in the literature with $A_{n,j}(q,t)$ expressed in terms of the derivatives of $P_n(q,t)$ which gives by Lemma 15

(38)
$$\sum_{j=0}^{n} \frac{(n-j)!\, N(q-1,j)}{\Gamma(n+j+q-2)} (1-t^2)^{j/2}(1-s^2)^{j/2} P_n^{(j)}(q,t)\, P_n^{(j)}(q,s)\, P_j(q-1, \xi_{q-1}\cdot\eta_{q-1}) =$$

$$= \frac{n!}{\Gamma(n+q-2)} P_n(q, t\cdot s + \sqrt{1-t^2}\cdot\sqrt{1-s^2}\,(\xi_{q-1}\cdot\eta_{q-1})) .$$

PROPERTIES OF THE LEGENDRE FUNCTIONS

Multiplying (37) by $P_\ell(q-1, \xi_{q-1}\cdot\eta_{q-1})$ and integrating over Ω_{q-1} with respect to ξ_{q-1}, we get from 23 and Lemma (10) :

$$A_{n,\ell}(q,t)\, A_{n,\ell}(q,s) = \frac{N(q,n)\,\omega_{q-2}}{\omega_q} \int_{-1}^{+1} P_n(q, t\cdot s + \sqrt{1-t^2}\cdot\sqrt{1-s^2}\cdot v)\, P_\ell(q-1,v)(1-v^2)^{\frac{q-4}{2}} dv .$$

From Lemma 15 we now get

Lemma 16:

$$\frac{N(2\ell+q, n-\ell)\,\omega_{2\ell+q-1}}{\omega_{2\ell+q}} (1-t^2)^\ell P_{n-\ell}(2\ell+q,t)\, P_{n-\ell}(2\ell+q,s)$$

$$= \frac{N(q,n)\,\omega_{q-2}}{\omega_q} \int_{-1}^{+1} P_n(q, t\cdot s + \sqrt{1-t^2}\cdot\sqrt{1-s^2}\cdot v)\, P_\ell(q-1,v)(1-v^2)^{\frac{q-4}{2}} dv.$$

In particular it follows for $\ell = 0$

$$\frac{\omega_{q-1}}{\omega_{q-2}} P_n(q,t)\, P_n(q,s) = \int_{-1}^{+1} P_n(q, t\cdot s + \sqrt{1-t^2}\sqrt{1-s^2}\cdot v)(1-v^2)^{\frac{q-4}{2}} dv .$$

We now prove

Lemma 17 : For $0 \leq x < 1$ and $-1 \leq t \leq 1$,

$$\sum_{n=0}^{\infty} N(q,n)\, x^n P_n(q,t) = \frac{1-x^2}{(1+x^2-2xt)^{q/2}} .$$

For $q = 2$ this is a well-known identity which we can best obtain by setting $t = \cos\varphi$. Then

$$\sum_{n=0}^{\infty} N(2,n)\, x^n P_n(2,t) = \sum_{n=-\infty}^{+\infty} x^{|n|} e^{in\varphi} = \frac{1}{1-xe^{i\varphi}} + \frac{1}{1-xe^{-i\varphi}} - 1$$

$$= \frac{2-2x\cos\varphi}{1+x^2-2x\cos\varphi} - 1 = \frac{1-x^2}{1+x^2-2x\cos\varphi}$$

$$= \frac{1-x^2}{1+x^2-2xt} .$$

We may assume for the following, therefore, that $q \geq 3$.

Using the Laplace representation (Theorem 7) of the Legendre polynomials, we find for the left hand side

$$(39) \qquad \frac{\omega_{q-2}}{\omega_{q-1}} \sum_{n=0}^{\infty} N(q,n) \int_{-1}^{+1} x^n (t + i\sqrt{1-t^2}\cdot s)^n (1-s^2)^{\frac{q-4}{2}}\, ds .$$

In Lemma 3, we had proved the identity

$$(40) \qquad \sum_{n=0}^{\infty} N(q,n)\, x^n = \frac{1+x}{(1-x)^{q-1}} .$$

As $|t+i\sqrt{1-t^2}\, s|^2 = t^2+(1-t^2)s^2 \leq t^2+(1-t^2) = 1$ and hence $|x(t+i\sqrt{1-t^2}\, s)|$ is less than one we may write - under the condition stated in Lemma 17 - the formula (39) as

$$(41) \qquad \frac{\omega_{q-2}}{\omega_{q-1}} \int_{-1}^{+1} \frac{1+x(t+i\sqrt{1-t^2}\cdot s)}{[1-x(t+i\sqrt{1-t^2}\cdot s)]^{q-1}} (1-s^2)^{\frac{q-4}{2}}\, ds .$$

To prove our Lemma, we thus have to show that this integral is equal to the function given on the right hand side of Lemma 17. In order to do this we introduce the substitution s = tanh u.

Using the abbreviations

$$f_1(u) = (1+xt)\cosh u + ix\sqrt{1-t^2}\,\sinh u \tag{42}$$
$$f_2(u) = (1-xt)\cosh u - ix\sqrt{1-t^2}\,\sinh u$$

and observing

$$(1-s^2)^{\frac{q-4}{2}}\,ds = (\cosh u)^{2-q}\,du$$

we obtain for the integral in (41)

$$\int_{-\infty}^{+\infty} \frac{f_1(u)}{[f_2(u)]^{q-1}}\,du\,. \tag{43}$$

If f(u) stands for either of the two functions defined in (42), we have from $f''(u) = +f(u)$ for any complex number u_0

$$f(u) = f(u_0)\cosh(u-u_0) + f'(u_0)\sinh(u-u_0).$$

Now we introduce the real number γ by

$$x\sqrt{1-t^2} + i(1-xt) = \sqrt{1+x^2-2xt}\;e^{i\gamma}\,,\quad 0<\gamma<\frac{\pi}{2} \tag{44}$$

so that we can write $f_2(u)$ as

$$f_2(u) = -i\sqrt{1+x^2-2xt}\;\sinh(u+i\gamma)\,. \tag{45}$$

Apart from a numerical constant the integral (43) therefore equals

$$\frac{i^{q-1}}{(1+x^2-2xt)^{\frac{q-1}{2}}}\int_{-\infty}^{+\infty}\frac{f_1(-i\gamma)\cosh(u+i\gamma)+f_1'(-i\gamma)\sinh(u+i\gamma)}{[\sinh(u+i\gamma)]^{q-1}}\,du\,.$$

Here the integral reduces to

$$f_1'(-i\gamma)\int_{-\infty}^{+\infty}\frac{du}{[\sinh(u+i\gamma)]^{q-2}} + f_1(-i\gamma)\int_{-\infty}^{+\infty}\frac{\cosh(u+i\gamma)}{[\sinh(u+i\gamma)]^{q-1}}\,du \tag{46}$$

where the second term of this sum vanishes, because the integral is zero. As γ is greater than zero, the integral in (46) exists for all $q \geq 3$. It may be regarded as a complex integral

$$\int_{-\infty+i\gamma}^{+\infty+i\gamma}\frac{du}{(\sinh u)^{q-2}} = \int_{-\infty+i\pi/2}^{+\infty+i\pi/2}\frac{du}{(\sinh u)^{q-2}} \tag{47}$$

where this last identity is obtained by shifting the path of integration to the line $\mathrm{Im}(u) = \frac{\pi}{2}$. Combining these results and expressing the integral in (47) by use of the substitution $u = v + i\frac{\pi}{2}$, we now see that

$$\sum_{n=0}^{\infty} N(q,n)\, x^n P_n(q,t) = \frac{i\, f_1'(-i\gamma)}{(1+x^2-2xt)^{\frac{q-1}{2}}} \frac{\omega_{q-2}}{\omega_{q-1}} \int_{-\infty}^{+\infty} \frac{dv}{(\cosh v)^{q-2}} .$$

From (42) and (44) we get

$$i f_1'(-i\gamma) = (1+xt)\sin\gamma - x\sqrt{1-t^2}\cos\gamma = \frac{1-x^2}{\sqrt{1+x^2-2xt}} .$$

We have thus proved that

(48)
$$\sum_{n=0}^{\infty} N(q,n)\, x^n P_n(q,t) = C\, \frac{1-x^2}{(1+x^2-2xt)^{q/2}} .$$

In order to determine the constant C we set t = 1, and obtain from (40)

$$\sum_{n=0}^{\infty} N(q,n)\, x^n = \frac{1+x}{(1-x)^{q-1}} .$$

As the right hand side of (48) reduces to this value for t = 1, we obtain C = 1 and have thus proved our identity.

Introducing

$$S_n(q,t) = \sum_{k=0}^{n} N(q,k)\, P_k(q,t) \;;\quad S_0(q,t) = 1$$

we have

$$\sum_{n=0}^{\infty} N(q,n)\, x^n P_n(q,t) = 1 + \sum_{n=1}^{\infty} x^n \left(S_n(q,t) - S_{n-1}(q,t) \right)$$

$$= \sum_{n=0}^{\infty} x^n S_n(q,t) - x \sum_{n=0}^{\infty} x^n S_n(q,t)$$

where

(49)
$$\sum_{n=0}^{\infty} x^n S_n(q,t) = \frac{1+x}{(1+x^2-2xt)^{q/2}} .$$

Set for n = 0,1,..., and q ≥ 3,

$$c_n(q) = (-1)^n \binom{2-q}{n} = \frac{\Gamma(n+q-2)}{\Gamma(n+1)\cdot\Gamma(q-2)} ,$$

so that

$$\sum_{n=0}^{\infty} c_n(q)\, x^n = \frac{1}{(1-x)^{q-2}} \quad .$$

We then get

Lemma 18 : For $q \geq 3$, $0 \leq x < 1$, and $-1 \leq t \leq 1$

$$\sum_{n=0}^{\infty} c_n(q)\, x^n\, P_n(q,t) = \frac{1}{(1+x^2-2xt)^{\frac{q-2}{2}}} \qquad {}^{1)}$$

with

$$c_n(q) = \frac{\Gamma(n+q-2)}{\Gamma(q-2)\cdot\Gamma(n+1)} \quad ;$$

The corresponding result for $q = 2$ is

$$\sum_{n=1}^{\infty} \frac{x^n}{n}\, P_n(2,t) = \frac{1}{2}\ln(1+x^2-2xt)$$

which is well known and can be proved immediately by using $P_n(2,\cos\varphi) = \cos n\varphi$.

The proof of Lemma 18 is quite analogous to the proof of Lemma 17 so that we can use the same notations. Laplace's representation of $P_n(q,t)$ gives

$$\sum_{n=0}^{\infty} c_n(q)\, x^n\, P_n(q,t) = \frac{\omega_{q-2}}{\omega_{q-1}} \int_{-1}^{+1} \frac{(1-s^2)^{\frac{q-4}{2}}\, ds}{(1-x(t+i\sqrt{1-t^2}\cdot s))^{q-2}} \quad .$$

The substitution $s = \tanh u$ and the abbreviations (42), (44), (45) transform the integral to

$$\int_{-1}^{+1} \frac{du}{[f_2(u)]^{q-2}} = \frac{(i)^{q-2}}{(1+x^2-2xt)^{\frac{q-2}{2}}} \cdot \int_{-\infty}^{+\infty} \frac{du}{[\sinh(u+i\gamma)]^{q-2}}$$

so that Lemma 18 may be proved by the same arguments that led to Lemma 17.

---

1) This estabilishes the relation $c_n(q) P_n(q,t) = C_n^{(q-2)/2}(t)$ where $C_n^{\nu}(t)$ are the Gegenbauer functions.

Suppose now that x and y are any two vectors in q-dimensional space, with

$$x = R\cdot\xi \, , \quad y = r\cdot\eta \; ; \quad |\xi| = 1 \, , \quad |\eta| = 1 \, .$$

Then for $q \geqq 3$ and $R > r$

$$\frac{1}{|x-y|^{q-2}} = \frac{1}{(R^2 + r^2 - 2R\cdot r\, \xi\cdot\eta)^{\frac{q-2}{2}}} =$$

$$= \frac{1}{R^{q-2}} \, \frac{1}{\left(1 + \left(\frac{r}{R}\right)^2 - 2\left(\frac{r}{R}\right)\xi\cdot\eta\right)^{\frac{q-2}{2}}} \, .$$

This can be expressed by Lemma 18 so that we obtain

Lemma 19: If $x = R\,\xi$; $y = r\cdot\eta$ and $R > r$, then

$$|x-y|^{2-q} = R^{2-q} \sum_{n=0}^{\infty} C_n(q) \left(\frac{r}{R}\right)^n P_n(q, \xi\cdot\eta) \, .$$

Let x_i, y_i be the Cartesian components of x and y. Then

(50) $$|x-y|^{2-q} = \left[(x_1-y_1)^2 + (x_2-y_2)^2 + \cdots + (x_q-y_q)^2 \right]^{\frac{2-q}{2}} .$$

According to the Taylor expansion in several variables, this can be written as

(51) $$\sum_{n=0}^{\infty} \frac{(-1)^n}{n!} \left(y_1 \frac{\partial}{\partial x_1} + \cdots\cdots + y_q \frac{\partial}{\partial x_q} \right)^n \left[(x_1)^2 + \cdots\cdots + (x_q)^2 \right]^{\frac{2-q}{2}} .$$

If ∇_x denotes as usual the vector operator with the components $\frac{\partial}{\partial x_i}$ we have

$$y_1 \frac{\partial}{\partial x_1} + \cdots\cdots + y_q \frac{\partial}{\partial x_q} = r\,(\eta\cdot\nabla_x)$$

and we get from (50) and (51) for $|y| < |x|$

$$|x-y|^{2-q} = \sum_{n=0}^{\infty} \frac{(-1)^n}{n!} r^n (\eta\cdot\nabla_x)^n \, |x|^{2-q} \, .$$

Comparing this with Lemma 19 we have by equating the coefficients of r^n,

$$\frac{(-1)^n}{n!}(\eta \cdot \nabla_x)^n |x|^{2-q} = c_n(q) \frac{P_n(q, \xi \cdot \eta)}{R^{n+q-2}} .$$

This gives with the explicit value of $c_n(q)$ and $R = |x|$

Lemma 20 : (Maxwell's representation)

$$(\eta \cdot \nabla_x)^n |x|^{2-q} = (-1)^n \frac{\Gamma(n+q-2)}{\Gamma(q-2)} \cdot \frac{P_n(q, \xi \cdot \eta)}{|x|^{n+q-2}} .$$

As $|x|^{2-q}$ is the fundamental solution of the Laplace equation in q dimensions, this shows that the Legendre polynomials may be obtained by repeated differentiations of the fundamental solution in the direction of the vector η. The potential on the right hand side of Lemma 20 may thus be regarded as the potential of a pole of order n with the axis η at the origin.

We know that every spherical harmonic can be expressed in the form

$$S_n(q, \xi) = \sum_{k=1}^{N(q,n)} a_k P_n(q, \xi \cdot \eta_k)$$

with a fundamental system η_k. Therefore it is always possible to write

$$\frac{S_n(q, \xi)}{r^{n+q-2}} = (-1)^n \cdot \frac{\Gamma(q-2)}{\Gamma(n+q-2)} \sum_{k=1}^{N(q,n)} a_k (\eta_k \cdot \nabla_x)^n |x|^{2-q}$$

which shows that every potential of this type may be regarded as the potential of a combination of multipoles with real axis. The system of fundamental points introduced earlier thus corresponds to a fundamental system of multipoles in Maxwell's interpretation of the spherical harmonics.

A rather striking interpretation of Lemma 20 is obtained in the following way. We first observe that

$$\int_{\Omega_q} (x \cdot \eta)^n S_n(q, \eta) \, d\omega_q(\eta) = \lambda_n H_n(q, x)$$

with

$$\lambda_n = \omega_{q-1} \int_{-1}^{+1} t^n P_n(q,t)(1-t^2)^{\frac{q-3}{2}} dt$$

$$= \sqrt{\pi}\cdot\omega_{q-1}\cdot\Gamma\left(\tfrac{q-1}{2}\right)\cdot 2^{-n}\frac{\Gamma(n+1)}{\Gamma(n+\frac{q}{2})}$$

where $H_n(q,x) = r^n S_n(q,\xi)$, which enables us to express the formal polynomial $H_n(q,\nabla_x)$ as

$$\lambda_n H_n(q,\nabla_x) = \int_{\Omega_q} S_n(q,\eta)\,(\eta\cdot\nabla_x)^n\, d\omega_q(\eta)\,.$$

Multiplication of both sides of Lemma 20 with $S_n(q,\eta)$ and integration over Ω_q now gives

Lemma 21 : For every harmonic polynomial of degree n

$$H_n(q,\nabla_x)\,|x|^{2-q} = (-1)^n \frac{2^n\,\Gamma\left(\frac{2n+q-2}{2}\right)}{\Gamma\left(\frac{q-2}{2}\right)}\cdot\frac{H_n(q,x)}{|x|^{2n+q-2}}\,.$$

Before leaving the special properties of the spherical harmonics it should be noted that many more can be derived from Lemmas 18 to 21 of which the recursion formulas for the Legendre polynomials, the associated functions, and their derivatives are perhaps best known. They can be obtained by differentiating the identity formulated in Lemma 18 with respect to x or t and equating coefficients of x^n.

As an example we take the formula

$$(52)\qquad (q-2)\cdot\sum_{k=0}^{n} N(q,k)\,P_k(q,t) = c_n(q)\,P_n'(q,t) + c_{n+1}(q)\,P_{n+1}'(q,t)\,.$$

From the Laplace representation (Theorem 7) we get

$$P_n'(q,t) = \frac{\omega_{q-2}}{\omega_{q-1}}\int_{-1}^{+1} n\,(t+i\sqrt{1-t^2}\cdot s)^{n-1}\left(1-i\cdot\frac{t}{\sqrt{1-t^2}}\,s\right)(1-s^2)^{\frac{q-4}{2}}\,ds,$$

which shows that for all t with $|t| \le t_0 < 1$, $P_n'(q,t)$ satisfies

$$P_n'(q,t) = \mathcal{O}(n)$$

uniformly. It is therefore permitted to differentiate the power series of Lemma 18 termwise. We obtain

$$\sum_{n=0}^{\infty} c_n(q)\, x^n\, P_n'(q,t) = (q-2)\,\frac{x}{(1+x^2-2xt)^{q/2}}$$

which gives us

$$\sum_{n=0}^{\infty} x^n \left[\, c_n(q)\, P_n'(q,t) + c_{n+1}(q)\, P_{n+1}'(q,t)\right] = (q-2)\,\frac{1+x}{(1+x^2-2xt)^{q/2}} .$$

Comparing this result with (49) and equating coefficients of x^n we get (52). This becomes particularly simple for $q = 3$, as is true of many more of these results. In this case we get

$$\sum_{k=0}^{n} (2k+1)\, P_k(3,t) = P_n'(3,t) + P_{n+1}'(3,t) .$$

DIFFERENTIAL EQUATIONS

The basic concept and the starting point of our approach to the theory of spherical harmonics is the harmonic and homogeneous polynomial. Only very indirectly we made use of the fact that the spherical harmonics are connected with the Laplace equation. We shall now derive results which express this factor in terms of special differential equations for the spherical harmonics.

In order to do this we have to express the Δ-operator in terms of the polar coordinates which we have been using. We wrote

$$x = r\,(t\,\varepsilon_q + \sqrt{1-t^2}\;\xi_{q-1}) \tag{53}$$

where ξ_{q-1} is a unit vector spanned by the unit vectors $\varepsilon_1, \ldots\ldots, \varepsilon_{q-1}$. Suppose now that we have some coordinate representation $v_1, \ldots\ldots, v_{q-1}$ of Ω_{q-1}. We then set

$$u_q = r \;;\quad u_{q-1} = t \;;\quad u_i = v_i \quad \text{for} \quad i = 1, \ldots, q-2$$

so that ξ_q is a function of t and $v_1, \ldots\ldots, v_{q-1}$, or in the above notation of $u_1, \ldots\ldots, u_{q-1}$. With the abbreviation

$$g_{ik} = \frac{\partial \xi_q}{\partial u_i} \cdot \frac{\partial \xi_q}{\partial u_k} \; ; \; g = \det(g_{ik}) \; ; \; g^{ik} g_{ij} = \delta_j^k \, , \quad i,k = 1,2,\dots,q-1;$$

we may form the Beltrami Operator for Ω_q

$$\Delta_q^* = \frac{1}{\sqrt{g}} \sum_{i=1}^{q-1} \sum_{k=1}^{q-1} \frac{\partial}{\partial u_i} \sqrt{g} \, g^{ik} \frac{\partial}{\partial u_k} \, .$$

From (53) it is clear that for $i,k = 1,2\dots,q-1$,

$$\frac{\partial x}{\partial u_q} = \xi_q \; ; \; \frac{\partial x}{\partial u_i} = r \cdot \frac{\partial \xi_q}{\partial u_i}$$

$$\frac{\partial x}{\partial u_i} \cdot \frac{\partial x}{\partial u_k} = r^2 g_{ik} \; ; \; \frac{\partial x}{\partial u_i} \cdot \frac{\partial x}{\partial u_q} = 0 \; ; \; \frac{\partial x}{\partial u_q} \frac{\partial x}{\partial u_q} = 1$$

and we obtain by means of the tensor calculus

$$\Delta_q = \frac{\partial^2}{\partial r^2} + (q-1) \frac{1}{r} \frac{\partial}{\partial r} + \frac{1}{r^2} \Delta_q^* \, .$$

We had

$$\xi_q = t \cdot \varepsilon_q + \sqrt{1-t^2} \cdot \xi_{q-1}$$

so that for $i,k = 1,\dots,q-2$,

$$\frac{\partial \xi_q}{\partial u_{q-1}} \cdot \frac{\partial \xi_q}{\partial u_{q-1}} = \frac{1}{1-t^2} \; ; \qquad \frac{\partial \xi_q}{\partial u_{q-1}} \cdot \frac{\partial \xi_q}{\partial u_i} = 0 \; ;$$

$$\frac{\partial \xi_q}{\partial u_i} \cdot \frac{\partial \xi_q}{\partial u_k} = \frac{\partial \xi_{q-1}}{\partial u_i} \cdot \frac{\partial \xi_{q-1}}{\partial u_k} \cdot (1-t^2) \, .$$

This gives us

(54) $$\Delta_q^* = (1-t^2) \frac{\partial^2}{\partial t^2} - (q-1) t \frac{\partial}{\partial t} + \frac{1}{1-t^2} \Delta_{q-1}^* \, .$$

It should be noted that for

$$\xi_2 = (\cos \varphi) \cdot \varepsilon_1 + (\sin \varphi) \cdot \varepsilon_2$$

we get

$$\Delta_2^* = \frac{\partial^2}{\partial \varphi^2} \, .$$

We can thus define the operators Δ_q^* successively, starting with the two -dimensional case.

As $r^n S_n(q, \xi)$ is a harmonic function we get

$$0 = \Delta_q r^n S_n(q,\xi) = n(n+q-2) r^{n-2} S_n(q,\xi) + r^{n-2} \Delta_q^* S_n(q,\xi)$$

which gives us

<u>Lemma 22</u> : Every spherical harmonic of degree n and dimension q satisfies

$$\Delta_q^* S_n(q,\xi) + n(n+q-2) S_n(q,\xi) = 0 .$$

For the Legendre polynomials we thus get from (54)

<u>Lemma 23</u> : The Legendre polynomial $P_n(q,t)$ satisfies

$$[(1-t^2)\frac{d^2}{dt^2} - (q-1)t\frac{d}{dt}] P_n(q,t) + n(n+q-2) P_n(q,t) = 0.$$

The associated Legendre functions satisfy

$$[\Delta_q^* + n(n+q-2)] A_{n,j}(q,t) S_j(q-1,\xi_{q-1}) = 0$$

which gives us

<u>Lemma 24</u> : The associated Legendre functions $A_{n,j}(q,t)$ of degree n, order j, and dimension q satisfy

$$[(1-t^2)\frac{d^2}{dt^2} - (q-1)t\frac{d}{dt} + n(n+q-2) - \frac{j(j+q-3)}{1-t^2}] A_{n,j}(q,t) = 0.$$

The extension of the concept of spherical harmonics for degrees and orders which are not integers, may be started from these differential equations, as has been done previously (see Hobson, Spherical Harmonics). However, if the condition is imposed that the harmonic functions thus obtained should be entire and univalued, the theory reduces to the functions discussed here, which are therefore called the regular spherical harmonics.

EXPANSIONS IN SPHERICAL HARMONICS

We shall now prove that the spherical harmonics form a complete and closed set of functions on the sphere. This, of course, may be regarded as an extension of the theory of Fourier series to the case of problems with spherical symmetry in any number of dimensions.

Due to the orthogonality of the Legendre polynomials we have from Lemma 17 (multiply by P_0 and integrate)

$$(55) \qquad \int_{-1}^{+1} \frac{(1-x^2)(1-t^2)^{\frac{q-3}{2}}}{(1+x^2-2xt)^{q/2}}\,dt = \int_{-1}^{+1} (1-t^2)^{\frac{q-3}{2}}\,dt = \frac{\omega_q}{\omega_{q-1}}$$

for all x with $0 \leq x < 1$. We shall now prove

Lemma 25 : Suppose $f(t)$ is continuous for $-1 \leq t \leq 1$. Then

$$\lim_{x\to 1-0} \int_{-1}^{+1} \frac{(1-x^2)\,f(t)\,(1-t^2)^{\frac{q-3}{2}}}{(1+x^2-2xt)^{q/2}}\,dt = f(1)\cdot\frac{\omega_q}{\omega_{q-1}} .$$

We write

$$f(t) = f(1) + g(t)$$

where $g(1) = 0$. If $f(t)$ is constant the result follows from (55) immediately, so that Lemma 25 is proved if we show

$$\lim_{x\to 1-0} \int_{-1}^{+1} \frac{(1-x^2)\,g(t)\,(1-t^2)^{\frac{q-3}{2}}}{(1+x^2-2xt)^{q/2}}\,dt = 0 .$$

The continuity of $g(t)$ implies that there is a positive function $m(s)$ with

$$\lim_{s\to 0} m(s) = 0$$

such that

$$\max_{1\geq t\geq 1-s} |g(t)| \leq m(s) .$$

Moreover, it follows from the continuity that there is a constant C with

$$|g(t)| \leq C \quad \text{for} \quad -1 \leq t \leq 1 .$$

We now observe that for $-1 \leq t \leq 1-s$ and $x \geq 0$

$$1+x^2-2xt = (1-x)^2 + 2x(1-t) \geq 2x\cdot s$$

so that for the same range of t and $\frac{1}{2} < x < 1$

$$\frac{1-x^2}{(1+x^2-2xt)^{q/2}} \leq \frac{1-x^2}{(2x\cdot s)^{q/2}} = \frac{1+x}{(2x)^{q/2}}\cdot\frac{1-x}{s^{q/2}} \leq \frac{(1-x)\cdot 2}{s^{q/2}} .$$

We now define s by

(56) $$s^{q/2} = \sqrt{1-x}$$

and divide the interval of integration into $-1 \leq t \leq 1 - s$ and $1 - s \leq t \leq 1$. Then for $x \geq 0$

(57) $$\left| \int_{-1}^{1-s} \frac{(1-x^2)\, g(t)\, (1-t^2)^{\frac{q-3}{2}}}{(1+x^2-2xt)^{q/2}}\, dt \right| \leq 2C\,(1-x)^{1/2}\, \frac{\omega_q}{\omega_{q-1}} = O(\sqrt{1-x})$$

and

(58) $$\left| \int_{1-s}^{+1} \frac{(1-x^2)\, g(t)\, (1-t^2)^{\frac{q-3}{2}}}{(1+x^2-2xt)^{q/2}}\, dt \right| = O(m(s))$$

as this last integral may be majorized by

$$m(s) \int_{-1}^{+1} \frac{(1-x^2)\,(1-t^2)^{\frac{q-3}{2}}\, dt}{(1+x^2-2xt)^{q/2}} .$$

According to (55), s tends towards zero for $x \to 1 - 0$, so that our Lemma follows from (57) and (58) with (55).

We are now able to prove the following theorem:

<u>Theorem 8 :</u> (Poisson's integral) Suppose $F(\xi)$ is continuous on Ω_q . Then

$$\lim_{r \to 1-0} \frac{1}{\omega_q} \int_{\Omega_q} \frac{(1-r^2)\, F(\eta)}{(1+r^2-2r\,\xi\cdot\eta)^{q/2}}\, d\omega_q(\eta) = F(\xi)$$

where this limit holds uniformly with regard to ξ .

As Ω_q is compact, we can deduce from the continuity of $F(\xi)$ the existence of a positive function m(s) such that

(59) $$|F(\xi) - F(\eta)| \leq m(s) \quad \text{for} \quad 1 \geq \xi\cdot\eta \geq 1-s .$$

We now assume $\xi = \varepsilon_q$ and define

$$f(t) = \int_{\Omega_{q-1}} F\left(t\cdot\varepsilon_q + \sqrt{1-t^2}\,\eta_{q-1}\right) d\omega_{q-1}(\eta) ,$$

so that

$$f(1) = \omega_{q-1}\, F(\varepsilon_q) .$$

From (59) follows

(60) $$|f(1) - f(t)| \leq \omega_{q-1}\cdot m(s)$$

for $1 \geq t \geq 1 - s$. The integral in Theorem 8 can be written

$$\frac{1}{\omega_q}\int_{-1}^{+1}\frac{(1-\tau^2)\,f(t)\,(1-t^2)^{\frac{q-3}{2}}}{(1+\tau^2-2\tau t)^{q/2}}\,dt$$

so that we get for $\xi = \varepsilon_q$

$$\lim_{\tau\to 1-0}\frac{1}{\omega_q}\int_{\Omega_q}\frac{(1-\tau^2)\,F(\xi)}{(1+\tau^2-2\tau\cdot\xi\cdot\eta)^{q/2}}\,d\omega_q(\eta) =$$

$$\lim_{\tau\to 1-0}\frac{1}{\omega_q}\int_{-1}^{+1}\frac{(1-\tau^2)\,f(t)\,(1-t^2)^{\frac{q-3}{2}}}{(1+\tau^2-2\tau t)^{q/2}}\,dt =$$

$$= \frac{1}{\omega_q}\,\frac{\omega_q}{\omega_{q-1}}\,f(1) = F(\varepsilon_q)\,.$$

As any point of the sphere Ω_q may be chosen as ε_q of an appropriately chosen system of coordinates, this argument holds for all ξ of Ω_q. Moreover, the estimate (60) only involves the uniformly valid estimate (59) so that the limtis are approached uniformly.

From the identity

$$(61)\qquad \sum_{n=0}^{\infty} N(q,n)\,\tau^n\,P_n(q,\xi\cdot\eta) = \frac{1-\tau^2}{(1+\tau^2-2\tau\,\xi\cdot\eta)^{q/2}}$$

we may now deduce

<u>Theorem 9</u> : (Abel summation) Every function $F(\xi)$ which is continuous on Ω_q can be approximated uniformly in the sense of

$$\lim_{\tau\to 1-0}\sum_{n=0}^{\infty}\tau^n\,S_n(q,\xi) = F(\xi)$$

by spherical harmonics $S_n(q,\xi)$ which are given by

$$S_n(q,\xi) = \frac{N(q,n)}{\omega_q}\int_{\Omega_q} P_n(q,\xi\cdot\eta)\,F(\eta)\,d\omega_q(\eta) = \sum_{j=1}^{N(q,n)} c_{n,j}\,S_{n,j}(q,\xi)$$

where

$$c_{n,j} = \int_{\Omega_q} S_{n,j}(q,\eta)\, F(\eta)\, d\omega_q(\eta) .$$

This result is an immmediate consequence of the identity (61) which holds uniformly with respect to ξ and η for $0 \leq r < 1$. We may therefore integrate termwise and obtain the last representation of the spherical harmonics from the addition theorem.

Using the same notation, we get from Parseval's inequality

$$(62) \qquad \int_{\Omega_q} |F(\eta)|^2 d\omega_q(\eta) \geqslant \sum_{n=0}^{\infty} \sum_{j=1}^{N(q,n)} |c_{n,j}|^2 = \sum_{n=0}^{\infty} (c_n)^2 ,$$

where we used the abbreviation

$$\sum_{j=1}^{N(q,n)} |c_{n,j}|^2 = \int_{\Omega_q} |S_n(q,\eta)|^2 d\omega_q(\eta) = c_n^2 .$$

Set

$$F(r,\eta) = \sum_{n=0}^{\infty} r^n S_n(q,\eta) ; \quad F(1,\eta) = F(\eta) ,$$

so that

$$\lim_{r \to 1-0} \int_{\Omega_q} |F(r,\eta)|^2 d\omega_q(\eta) = \int_{\Omega_q} |F(\eta)|^2 d\omega_q(\eta)$$

as $F(r, \eta)$ approximates $F(\eta)$ uniformly. We therefore have

$$\lim_{r \to 1-0} \sum_{n=0}^{\infty} r^{2n} (c_n)^2 = \int_{\Omega_q} |F(\eta)|^2 d\omega_q(\eta) .$$

On the left hand side we may interchange the limit and the summation because of (62).

<u>Theorem 10</u> : For every continuous function $F(\xi)$

$$\int_{\Omega_q} |F(\xi)|^2 d\omega_q(\eta) = \sum_{n=0}^{\infty} (c_n)^2 .$$

Another conclusion may be drawn from Theorem 9.

<u>Theorem 11</u> : If the continuous function $F(\xi)$ satisfies

$$\int_{\Omega_q} F(\xi)\, S_n(q,\xi)\, d\omega_q(\xi) = 0$$

for all spherical harmonics, it vanishes identically.

Our assumption has the consequence that $F(r, \eta)$ vanishes for all $r < 1$. Therefore

$$\lim_{r \to 1-0} F(r, \eta) = F(\eta) = 0 ,$$

which proves Theorem 11.

These last results show that the system of spherical harmonics has the basic property of being complete and closed for the continuous functions on Ω_q . Extensions of these results to more general classes of functions may be obtained by methods of the theory of approximations.

BIBLIOGRAPHY

The following are either solely devoted to the subject of spherical harmonics or contain detailed information on this subject.

Erdélyi, A., W.Magnus, F. Oberhettinger, and F. Tricomi, Higher transcendental functions, Vol. 1 and 2, New York, 1953.

Hobson, E.W. The theory of spherical and ellipsoidal harmonics, Cambridge, 1931.

Lense, J. Kugelfunktionen, Leipzig 1950.

Magnus, W. and F. Oberhettinger, Formulas and theorems for the functions of mathematical physics, New York, 1954.

Müller, C., Grundprobleme der mathematischen Theorie elektromagnetischer Schwingungen, Berlin, Heidelberg, Göttingen, 1957.

Morse, P. M., and H. Feshbach, Methods of theoretical physics, Vol. 1 and 2, New York, 1953.

Sansone, G. Orthogonal functions, New York, 1959.

Webster, A.G. - Szegö, G. Partielle Differentialgleichungen der mathematischen Physik, Leipzig, Berlin, 1930.